CHAPTERWISE
ELEMENTS OF THE
PERIODIC TABLE
FLASH CARDS

Tool to score more marks in competitive exam such as
**IIT JEE, GATE, CBSE, CET, ICSE, Chemistry GRE,
AP Chemistry, Oxford Chemistry**

CHAPTERWISE

This is a carefully thought and crafted product by **CHAPTERWISE.**

Administrative Office:
Apurvamangal Craftspeople Private Limited (OPC)
Plot no 90, Survey no 65/1, Raavinagar,
Pashan Soos Road,
Pune 411021 Maharashtra India

All disputes are subject matter to Pune (MH) jurisdiction only.

ISBN-13: 978-1539447726
(CreateSpace-Assigned)
ISBN-10: 1539447723

Statistics has shown that a minimum of ~25% questions asked in IIT JEE are based on knowledge of elements. While cracking IIT exam remains as the main focus, knowing elements also helps in other as important entrance exams. Let me tell you that these marks can be a piece of cake if you know your elements well. More marks obviously is better ranking. It also helps you in Degree College as well as post-graduation and getting a doctorate.

How are CHAPTERWISE periodic elements flash cards designed?

There's a flawless rule that left psychological brain research known as the "Pictorial Superiority Effect." People remember pictures better than words, especially over longer periods of time. It refers to the notion that concepts that are learned by viewing pictures are more easily and frequently recalled than are concepts that are learned by reading their written word form counterparts. Sadly for us, studies are brimming with words, not pictures, so retention of information is low.

The most important aspect of scoring well in an exam is retention and recall. If retention is not good, nor will be the recall will.

"Based on research into the Picture Superiority Effect, when we read text alone, we are likely to remember only 10 percent of the information 3 days later. If that information is presented to us as text combined with a relevant image, we are likely to remember 65 percent of the information 3 days later."

- John Medina, Brain Rules, 2008

Many studies have also shown that having the learning material in hand rather on screen increases retention as it involves more sensory organs other than eyes. **CHAPTERWISE** has taken care of these mere facts and designed these flash cards so that you retain and recollect seamlessly. All elements are colour coded according to metal, metalloids and non-metals. While the front of card focuses on name, symbol posterior of card focuses on important characteristics of element. You can cut along the scissor line and use it as flash card or just carry the book.

So, without adieu, let's start knowing elements better.

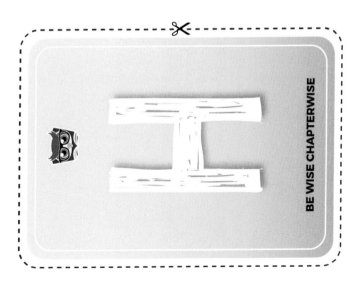

2 & 18

HELIUM

4.003

▲ 1s²

Nonmetals

noble gases

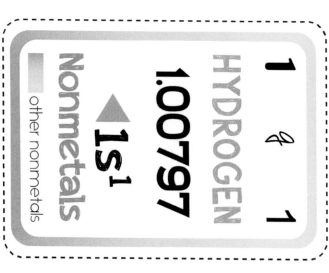

1 & 1

HYDROGEN

1.00797

▲ 1s¹

Nonmetals

other nonmetals

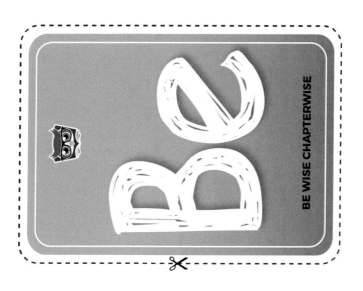

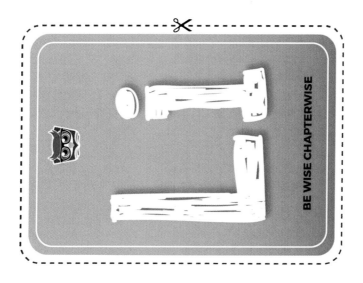

4

8   2

BERYLLIUM

9.102

▲ [He]2s²

Metals

alkaline earth metals

3

8   1

LITHIUM

6.939

▲ [He]2s¹

Metals

alkali metals

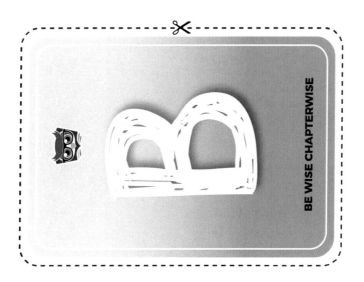

6

6   14

CARBON

12.011

▲ [He]2s² 2p²

Nonmetals

other nonmetals

5

5   13

BORON

10.811

▲ [He]2s²2p¹

Metals

metalloids

8

OXYGEN

15.9994

▲ [He]2s²2p⁴

Nonmetals

other nonmetals

16

7

NITROGEN

14.0067

▲ [He]2s²2p³

Nonmetals

other nonmetals

15

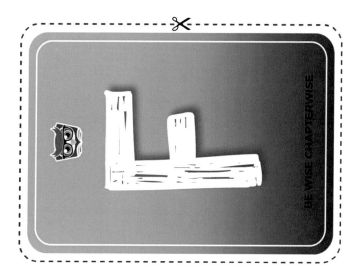

NEON

10

9

18

NEON

20.183

▲ [He]2s²2p⁶

Nonmetals

noble gases

FLUORINE

9

9

17

FLUORINE

18.993

▲ [He]2s²2p⁵

Nonmetals

halogens

BE WISE CHAPTERWISE

BE WISE CHAPTERWISE

12 s 2

MAGNESIUM

24.312

▲ [Ne]3s²

Metals

▮ alkaline earthmetals

---

11 s 1

SODIUM

22.9898

▲ [Ne]3s¹

Metals

▮ alkali metals

BE WISE CHAPTERWISE

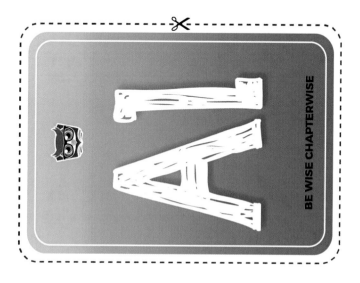

BE WISE CHAPTERWISE

14

S   14

SILICON

20.086

▲ [Ne]3s²3p²

Metals

metalloids

13

S   13

ALUMINIUM

26.982

▲ [Ne]3s²3p¹

Metals

post transition metals

16

S

16

SULFUR

32.064

▲ [Ne]3s²3p⁴

Nonmetals

other nonmetals

15

S

15

PHOSPHORUS

30.9738

▲ [Ne]3s²3p³

Nonmetals

other nonmetals

✂

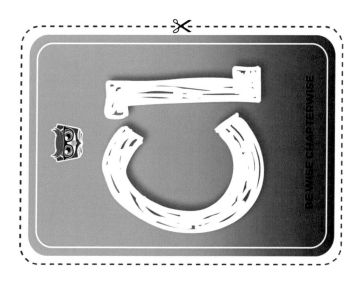

Card 1

18

Ar 18

ARGON

39.948

▲ [Ne]3s²3p⁶

Nonmetals

noble gases

Card 2

17

Cl 17

CHLORINE

35.453

▲ [Ne]3s²3p⁵

Nonmetals

halogens

BE WISE CHAPTERWISE

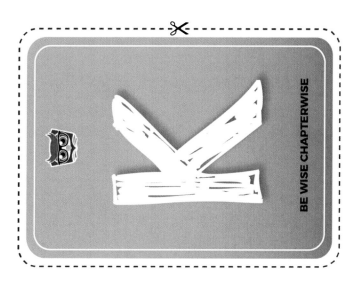

BE WISE CHAPTERWISE

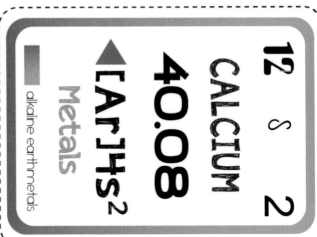

12

8 2

CALCIUM
40.08

▲ [Ar]4s²
Metals

alkaline earth metals

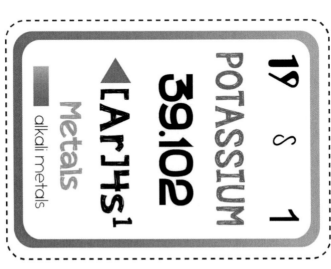

19

8 1

POTASSIUM
39.102

▲ [Ar]4s¹
Metals

alkali metals

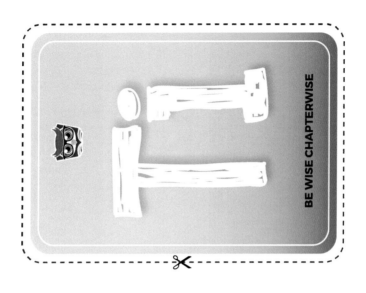

BE WISE CHAPTERWISE

BE WISE CHAPTERWISE

22 ⚛ 4

TITANIUM

47.9

▲ [Ar]3d²4s²

Metals

transition metals

---

21 ⚛ 3

SCANDIUM

44.96

▲ [Ar]3d¹4s²

Metals

transition metals

BE WISE CHAPTERWISE

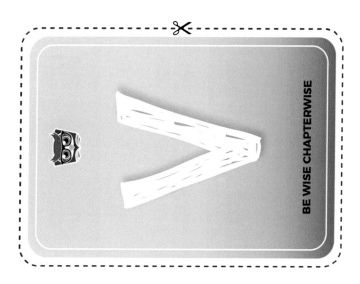

BE WISE CHAPTERWISE

24 S **6**

CHROMIUM

51.996

▲

[Ar]3d⁵4s¹

Metals

transition metals

---

23 S **5**

VANADIUM

50.942

▲

[Ar]3d³4s²

Metals

transition metals

BE WISE CHAPTERWISE

BE WISE CHAPTERWISE

26

IRON

55.847

$[Ar]3d^6 4s^2$

Metals

transition metals

8

25

MANGANESE

54.938

$[Ar]3d^5 4s^2$

Metals

transition metals

7

BE WISE CHAPTERWISE

BE WISE CHAPTERWISE

28 S 10

NICKEL

58.71

▲ [Ar]3d⁸4s²

Metals

transition metals

---

27 S 9

COBALT

58.933

▲ [Ar]3d⁷4s²

Metals

transition metals

BE WISE CHAPTERWISE

BE WISE CHAPTERWISE

30 8 **11**

ZINC

65.37

▲ [Ar]3d¹⁰4s²

Metals

■ transition metals

---

29 8 **11**

COPPER

63.54

▲ [Ar]3d¹⁰4s¹

Metals

■ transition metals

BE WISE CHAPTERWISE

BE WISE CHAPTERWISE

32   8   14

GERMANIUM

72.59

▲ $[Ar]3d^{10}4s^24p^2$

Metals

▬ metalloids

---

31   8   13

GALLIUM

69.72

▲ $[Ar]3s^{10}4s^24p^1$

Metals

▬ post transition metals

BE WISE CHAPTERWISE

BE WISE CHAPTERWISE

34

S 16

SELENIUM

78.96

▲[Ar]3d¹⁰4s²4p⁴

Nonmetals

other nonmetals

33

S 14

ARSENIC

74.992

▲[Ar]3d¹⁰4s²4p³

Metals

metalloids

BE WISE CHAPTERWISE

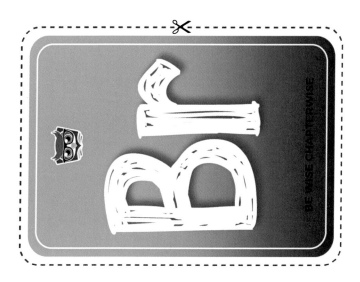

BE WISE CHAPTERWISE

36 Kr 18

KRYPTON
83.80

▲ [Ar]3d¹⁰4s²4p⁶

Nonmetals
noble gases

35 Br 17

BROMINE
79.909

▲ [Ar]3d¹⁰4s²4p⁵

Nonmetals
halogens

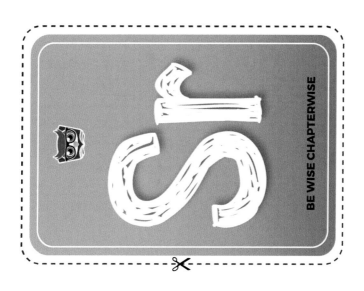

BE WISE CHAPTERWISE

BE WISE CHAPTERWISE

38   s   2

STRONTIUM

87.62

▲ [Kr]5s^2

Metals

■ alkaline earth metals

---

37   s   1

RUBIDIUM

85.47

▲ [Kr]5s^1

Metals

■ alkali metals

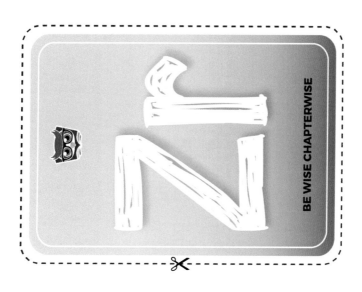

BE WISE CHAPTERWISE

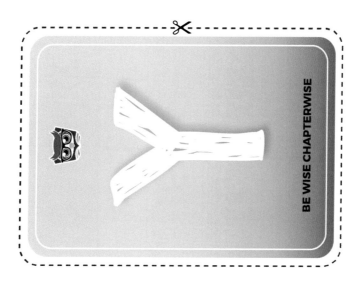

BE WISE CHAPTERWISE

40

S 4

ZIRCONIUM

91.22

▲[Kr]4d²5s²

Metals

transition metals

39

S 3

YTTRIUM

88.91

▲[Kr]4d¹5s²

Metals

transition metals

BE WISE CHAPTERWISE

BE WISE CHAPTERWISE

42 S 6

MOLYBDNUM

95.94

▲ [Kr]4d⁵5s¹

Metals

transition metals

41 S 5

NIOBIUM

92.91

▲ [Kr]4d⁴5s¹

Metals

transition metals

BE WISE CHAPTERWISE

BE WISE CHAPTERWISE

44   8   **8**

RUTHENIUM

186.22

▲ [Kr]4d^{7}5s^1

Metals

transition metals

---

43   8   **7**

TECHNETIUM

99

▲ [Kr]4d^{5}5s^2

Metals

transition metals

46   10

PALLADIUM

106.4

▲ [Kr]4d¹⁰

Metals

45   9

RHODIUM

102.91

▲ [Kr]4d⁸5s¹

Metals

BE WISE CHAPTERWISE

✂

BE WISE CHAPTERWISE

48 § 12

CADMIUM

112.40

▲ [Kr]4d¹⁰5s²

$[Kr]4d^{10}5s^2$

Metals

transition metals

---

47 § 11

SILVER

107.87

▲ [Kr]4d¹⁰5s¹

$[Kr]4d^{10}5s^1$

Metals

transition metals

BE WISE CHAPTERWISE

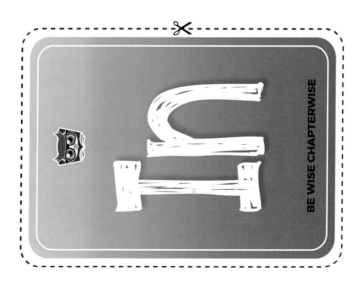

BE WISE CHAPTERWISE

50 Sn 14

TIN

118.69

[Kr]4d¹⁰ 5s² 5p²

Metals

post transition metals

49 In 13

INDIUM

114.82

[Kr]4d¹⁰ 5s² 5p¹

Metals

post transition metals

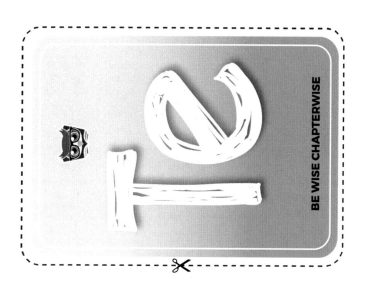

BE WISE CHAPTERWISE

BE WISE CHAPTERWISE

52 S 16

TELLURIUM

127.60

▲ [Kr]4d^{10}5s^{2}5p^4

Metals

metalloids

51 S 15

ANTIMONY

121.75

▲ [Kr]4d^{10}5s^{2}5p^3

Metals

metalloids

BE WISE CHAPTERWISE

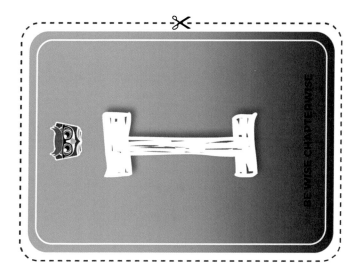

BE WISE CHAPTERWISE

54

⚛

18

XENON

131.30

▲

[Kr]4d¹⁰5s²5p⁶

Nonmetals

■ noble gases

53

§

7A

IODINE

126.904

▲

[Kr]4d¹⁰5s²5p⁵

Nonmetals

■ halogens

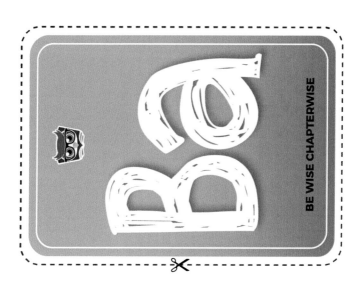

BE WISE CHAPTERWISE

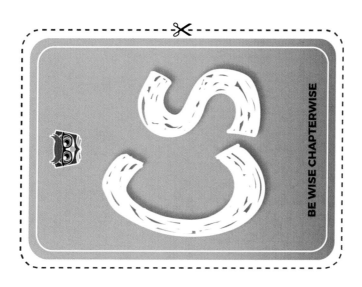

BE WISE CHAPTERWISE

56

8   2

BARIUM

137.34

▲ [Xe]6s²

Metals

■ alkaline earth metals

55

8   1

CAESIUM

132.905

▲ [Xe]6s¹

Metals

■ alkali metals

58

CERIUM

140.12

▲ [Xe]4f¹5d¹6s²

Metals

lanthanides

57

LANTHANUM

138.91

▲ [Xe]5d¹6s²

Metals

lanthanides

NO

BE WISE CHAPTERWISE

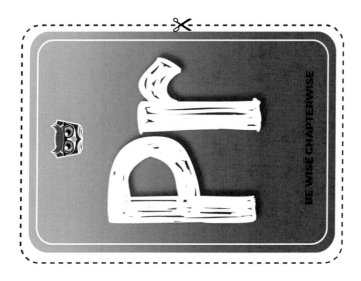

PI

BE WISE CHAPTERWISE

60 Nd

NEODYMIUM

144.24

▲ [Xe]4f⁴6s²

Metals

lanthanides

59 Pr

PRASEODYMIUM

140.91

▲ [Xe]4f³6s²

Metals

lanthanides

BE WISE CHAPTERWISE

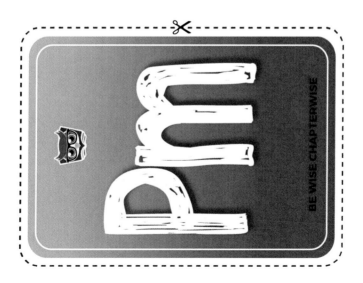

BE WISE CHAPTERWISE

62 Sm
SAMARIUM
150.35
▲ [Xe]4f⁶6s²
Metals
lanthanides

61 Pm
PROMETHIUM
147
▲ [Xe]4f⁵6s²
Metals
lanthanides

64 Gd
GADOLINIUM
157.25
▲ [Xe]4f⁷5d¹6s²
Metals
lanthanides

63 Eu
EUROPIUM
151.96
▲ [Xe]4f⁷6s²
Metals
lanthanides

66

Dy

DYSPROSIUM

162.50

▲ [Xe]4f¹⁰6s²

Metals

lanthanides

65

Tb

TERBIUM

153.92

▲ [Xe]4f⁹6s²

Metals

lanthanides

BE WISE CHAPTERWISE

BE WISE CHAPTERWISE

68

ERBIUM

167.26

▲ [Xe]4f¹²6s²

Metals

lanthanides

67

HOLMIUM

164.93

▲ [Xe]4f¹¹6s²

Metals

lanthanides

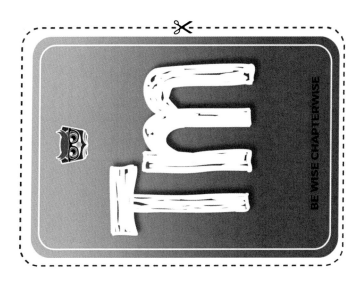

70

YTTERBIUM

173.04

▲ [Xe]4f¹⁴6s²

Metals

lanthanides

69

THULIUM

168.93

▲ [Xe]4f¹³6s²

Metals

lanthanides

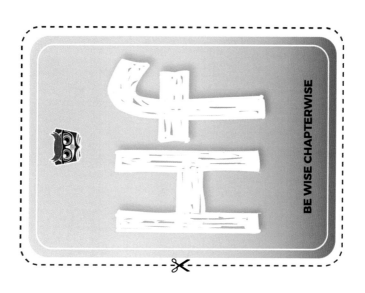

72   S   4

HAFNIUM

178.49

[Xe]4f¹⁴5d²6s²

Metals

transition metals

71   S

LUTETIUM

174.97

[Xe]4f¹⁴5d¹6s²

Metals

lanthanides

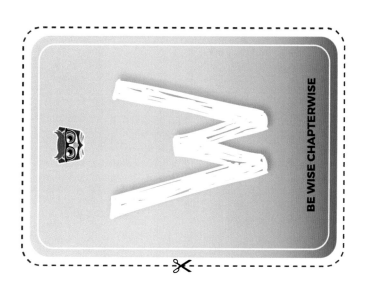

74 W 6

TUNGSTEN

183.85

▲ [Xe]4f¹⁴5d⁴6s²

Metals

transition metals

---

73 Ta 5

TANTALUM

180.95

▲ [Xe]4f¹⁴5d³6s²

Metals

transition metals

BE WISE CHAPTERWISE

BE WISE CHAPTERWISE

76

Os

8

OSMIUM

190.2

▲ [Xe]4f^{14}5d^{6}6s^2

Metals

transition metals

75

Re

7

RHENIUM

186.22

▲ [Xe]4f^{14}5d^{5}6s^2

Metals

transition metals

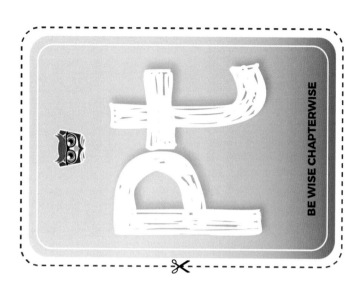

BE WISE CHAPTERWISE

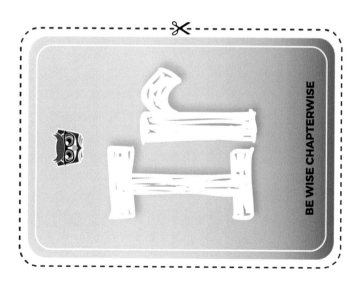

BE WISE CHAPTERWISE

78

PLATINUM

195.09

$[Xe]4f^{14}5d^9 6s^1$

Metals

transition metals

8 10

77

IRIDIUM

192.2

$[Xe]4f^{14}5d^7 6s^2$

Metals

transition metals

8 9

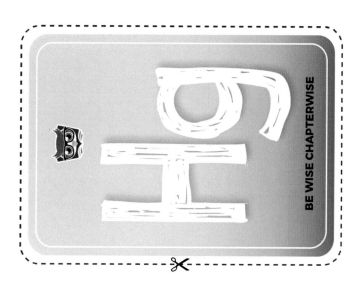

BE WISE CHAPTERWISE

BE WISE CHAPTERWISE

80   ℓ   12

MERCURY

200.59

▲ [Xe]4f^{14}5d^{10}6s^2

Metals

transition metals

79   δ   11

GOLD

196.967

▲ [Xe]4f^{14}5d^{10}6s^1

Metals

transition metals

BE WISE CHAPTERWISE

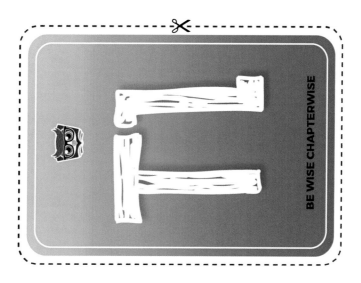

BE WISE CHAPTERWISE

82 § 14

LEAD

207.19

[Xe]4f¹⁴ 5d¹⁰ 6s² 6p²

$[Xe]4f^{14}\,5d^{10}\,6s^2\,6p^2$

Metals

post transition metals

---

81 § 13

THALLIUM

204.37

$[Xe]4f^{14}\,5d^{10}\,6s^2\,6p^1$

Metals

post transition metals

BE WISE CHAPTERWISE

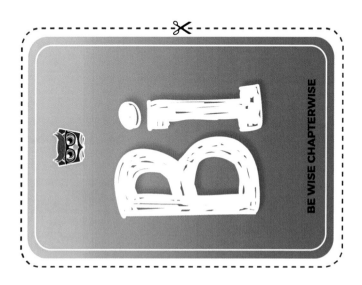

BE WISE CHAPTERWISE

84 § 16

POLONIUM

210

[Xe]4f^{14} 5d^{10} 6s^2 6p^4

Metals

metalloids

---

83 § 15

BISMUTH

208.98

[Xe]4f^{14} 5d^{10} 6s^2 6p^3

Metals

post transition metals

BE WISE CHAPTERWISE

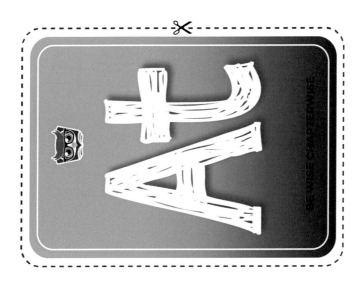

BE WISE CHAPTERWISE

86

Rn

18

RADON

222

▲

$[Xe]4f^{14}5d^{10}6s^26p^6$

Nonmetals

noble gases

85

At

17

ASTATINE

210

▲

$[Kr]4f^{14}5d^{10}6s^26p^5$

Nonmetals

halogens

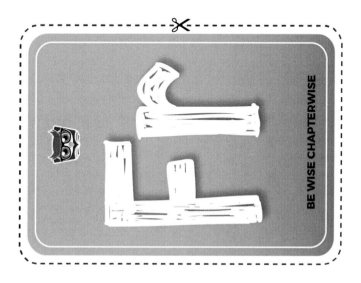

88 s 2

RADIUM

226

▲ [Rn]7s²

Metals

alkaline earth metals

87 s 1

FRANCIUM

223

▲ [Rn]7s¹

Metals

alkali metals

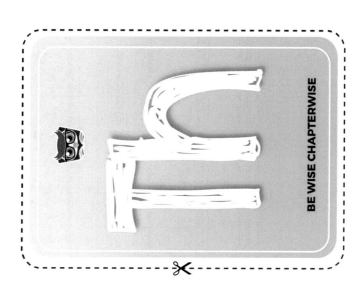

BE WISE CHAPTERWISE

BE WISE CHAPTERWISE

90

THORIUM

232.04

[Rn]6d²7s²

Metals

actinides

89

ACTINIUM

227

[Rn]6d¹7s²

Metals

actinides

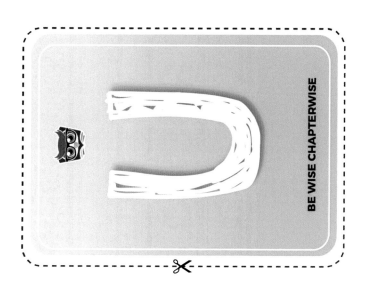

BE WISE CHAPTERWISE

BE WISE CHAPTERWISE

92 S

URANIUM

238.03

▲

[Rn]5f³6d¹7s²

Metals

actinides

---

91 S

PROTACTINIUM

231

▲

[Rn]5f²6d¹7s²

Metals

actinides

BE WISE CHAPTERWISE

BE WISE CHAPTERWISE

94   S

PLUTONIUM

242

$[Rn]5f^6 7s^2$

▲
Metals

█ actinides

---

93   S

NEPTUNIUM

237

$[Rn]5f^4 6d^1 7s^2$

▲
Metals

█ actinides

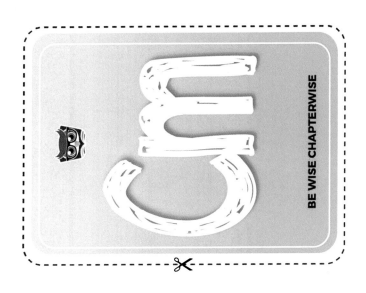

BE WISE CHAPTERWISE

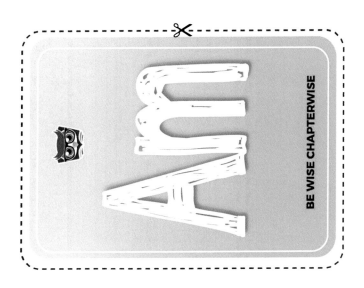

BE WISE CHAPTERWISE

96

CURIUM

247

[Rn]5f⁷6d¹7s²

Metals

actinides

95

AMERICIUM

243

[Rn]5f⁷7s²

Metals

actinides

BE WISE CHAPTERWISE

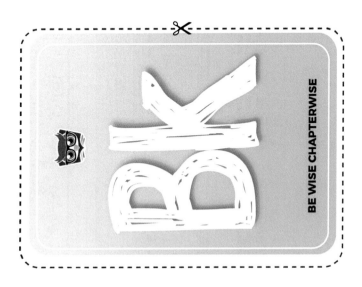

BE WISE CHAPTERWISE

98

S

CALIFORNIUM

251

▲ [Rn]5f¹⁰7s²

Metals

actinides

97

S

BERKELIUM

249

▲ [Rn]5f⁹7s²

Metals

actinides

BE WISE CHAPTERWISE

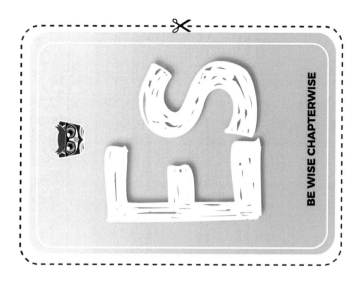

BE WISE CHAPTERWISE

100

s

FERMIUM

253

$[Rn]5f^{12}7s^2$

Metals

actinides

99

s

EINSTEINIUM

254

$[Rn]5f^{11}7s^2$

Metals

actinides

BE WISE CHAPTERWISE

BE WISE CHAPTERWISE

102 s

NOBELIUM

254

▲

[Rn]5f¹⁴7s²

Metals

actinides

---

101 s

MENDELEVIUM

256

▲

[Rn]5f¹³7s²

Metals

actinides

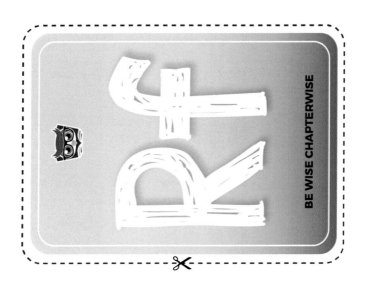

BE WISE CHAPTERWISE

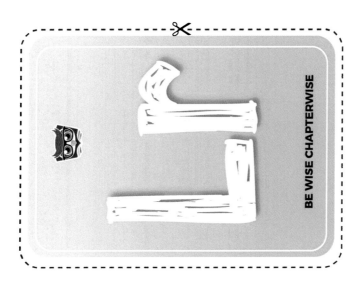

BE WISE CHAPTERWISE

104 s   4

RUTHERFORDIUM

261

[Rn]5f^{14}6d^{2}7s^2

Metals

transition metals

103 s

LAWRENCIUM

262

[Rn]5f^{14}7s^{2}7p^1

Metals

actinides

BE WISE CHAPTERWISE

BE WISE CHAPTERWISE

106 𝒔 6

SEABORGIUM

266

▲

$[Rn]5f^{14}6d^47s^2$

Metals

105 𝒔 5

DUBNIUM

262

▲

$[Rn]5f^{14}6d^37s^2$

Metals

BE WISE CHAPTERWISE

BE WISE CHAPTERWISE

108 s 8

HASSIUM

269

$[Rn]5f^{14}6d^67s^2$

▲ Metals

transition metals

107 s 7

BOHRIUM

264

$[Rn]5f^{14}6d^57s^2$

▲ Metals

transition metals

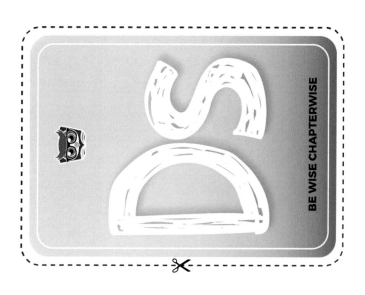

BE WISE CHAPTERWISE

BE WISE CHAPTERWISE

110

110 §§ 10

DARMSTADTIUM

281

[Rn]5f^{14}6d^{9}7s^1

Metals

transition metals

109

109 §§ 9

MEITNERIUM

268

[Rn]5f^{14}6d^{7}7s^2

Metals

transition metals

BE WISE CHAPTERWISE

BE WISE CHAPTERWISE

112 ℓ 12

COPERNICIUM

285

▲ [Rn]5f¹⁴6d¹⁰7s²

Metals
transition metals

---

111 s 11

ROENTGENIUM

280

▲ [Rn]5f¹⁴6d¹⁰7s¹

Metals
transition metals

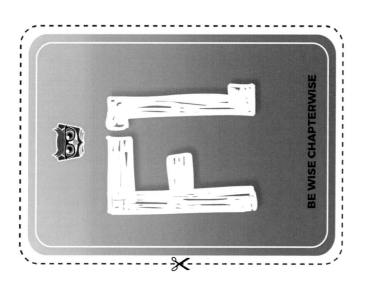

BE WISE CHAPTERWISE

BE WISE CHAPTERWISE

114   14

FLEROVIUM

289

▲

[Rn]5f^{14} 6d^{10} 7s^2 7p^2

Metals

post transition metals

---

113   13

UNUNTRIUM

286

▲

[Rn]5f^{14} 6d^{10} 7s^2 7p^1

Metals

post transition metals

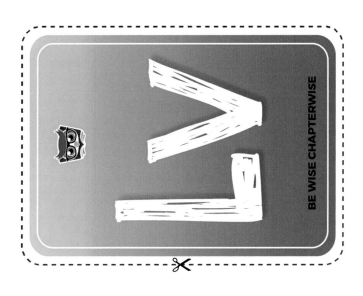

BE WISE CHAPTERWISE

BE WISE CHAPTERWISE

116 ∾ 16

LIVERMORIUM

293

▲ [Rn]5f¹⁴ 6d¹⁰ 7s² 7p⁴

Metals

post transition metals

115 ∾ 15

UNUNPENTIUM

289

▲ [Rn]5f¹⁴ 6d¹⁰ 7s² 7p³

Metals

post transition metals

uno

BE WISE CHAPTERWISE

uus

BE WISE CHAPTERWISE

118 ⚛ **18**

UNUNOCTIUM

294

▲ [Rn]5f^{14}6d^{10}7s^{2}7p^6

Nonmetals

◼ noble gases

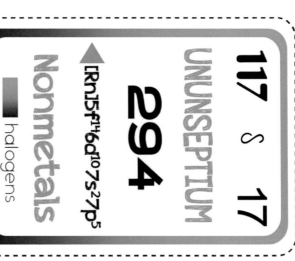

117 ⚛ **17**

UNUNSEPTIUM

294

▲ [Rn]5f^{14}6d^{10}7s^{2}7p^5

Nonmetals

◼ halogens

59887287R00071